To: ____________________

From: ________________

KODA and the WHALES

A TRUE STORY

Written by
Carrie Newell

Illustrated by
Jack Wiens

For information, address Top Reads Publishing Subsidiary Rights Department, 1035 E. Vista Way, Suite 205, Vista, CA 92084, USA.

ISBN: 978-1-61153-040-7 (hardcover)
ISBN: 978-1-61153-044-5 (paperback)
ISBN: 978-1-61153-045-2 (ebook)

Library of Congress Control Number: 2024917587

Koda and the Whales, a True Story is published by: Torchflame Books, an imprint of Top Reads Publishing, LLC, USA

For information about special discounts for bulk purchases,
please direct emails to: publisher@torchflamebooks.com

Cover design: Carrie Newell
Illustrations by: Jack Wiens
Book interior layout: Charles Hall and Carrie Newell

Printed in the United States of America

Dedication

This book is dedicated to my grandchildren: Kai and Kale whose dog is Lulu, Koda's sister, and Luke, Zach and Ellie who have Koda's other sister, Noodle, and to all dog lovers, especially children who love and have loved dogs and have had the opportunities to experience unique interactions between their dogs and other wildlife.

Left: Koda focused on watching whales
Middle: Kida almost nose to nose with a gray whale
Right: The Gray Whale, Blanco, checking out Kida

Acknowledgments

I would like to thank my incredible illustrator, Jack Wiens, my talented designer Charles Hall, my skillful editors Kim Wollenburg and Carrie Jenkins Williams and my awesome publisher Teri Rider with Torchflame Books for helping me get this book off the ground!

Other Books by Carrie Newell

Discovering Marine Biology along the Pacific Northwest Coast
A Guide to Summer Resident Gray Whales along the Oregon Coast
A Guide to Whales, Dolphins and Porpoises along the West Coast
A Guide to Seals, Sea Lions and Sea Otters along the West Coast
A Guide to Seabirds and Shorebirds along the Pacific Coast
Koda and the Whales: A True Story

KIDA
KODA
KILO

Koda and the Whales, A True Story

Author's Note…

This educational book teaches children facts about gray whales and their ocean home from the perspective of my dog, Koda, as a narrator. It is based on true stories and experiences I had with whales and all three of my dogs for decades. Many of the drawings are from photographs or a compilation of photographs with an artistic flair to describe Koda's story. Jack, my friend and illustrator for this book, is an incredible artist who is extremely accurate in his representations of marine life and my dogs' personalities.

For many years I have been blessed with three awesome dogs. Kida, a Golden retriever/ Malamute mix who passed in 2016 at the age of 15 years, Koda, a Golden doodle (Golden retriever/poodle mix) who was born in 2016, and our newest whale-finding dog, Kilo, who is Koda's daughter and who was born in December 2022. She is also a Golden doodle. I trained all of them as puppies to spot gray whales and alert me by barking. To train them, I positioned the boat and dogs downwind of the gray whales to let them smell the stinky breath and see and hear the whale's blow. At the moment the whale surfaced, I would enthusiastically say "Whale! Whale! Whale!" Soon the dogs understood that my excitement at sighting the whale was connected to their smelling of the whale's blow. No treats were involved as a reward, just lots of love and positive energy to motivate them to be alert for the whales.

Besides teaching about whales and other interesting facts, I am also demonstrating through my dogs and the whales that individuality, family, and friendships are important values. As I have experienced, losing a pet can be traumatic. I hope Koda's story will give children a sense that they can find peace in remembering the joys their animal friends brought to their lives.

The setting for this story is a quaint little town called Depoe Bay, nestled halfway up the Oregon Coast between the magnificent Coast Range Mountains and the mighty Pacific Ocean. It was named originally for Siletz Chief Charley DePoe. This beautiful town has spectacular red sunrises over the mountains and yellow and orange hued sunsets to the west, often with whale spouts blowing in the foreground. Depoe Bay's claim to fame is that it is "The Whale Watching Capital of the World." Plus, it is the smallest harbor in the world that big boats can get into and out of. The narrow opening can be extremely treacherous to cross during storms and high seas. But what is most special about Depoe Bay is that it is home to a group of summer resident whales who arrive to feed in the shallow coastal waters from May-October. These same whales come back every year and I get to know them as family.

KODA'S AUTOGRAPH PAGE

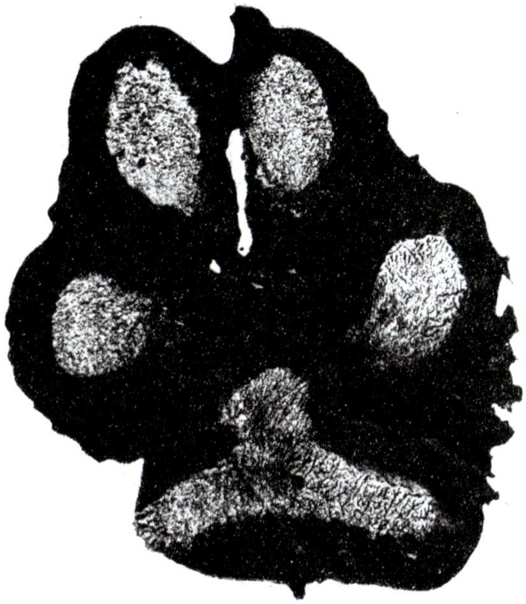

Koda

Follow Koda on her social media pages

Instagram
@Koda.Newell

Facebook
@Koda.Whale.Dog

When it All Started...

Hello, my name is Koda, and I am going to tell you the story of how my Mamma Carrie got me into finding whales for her whale-watching tours. My Mamma Carrie loves whales and especially gray whales. When I was a 12-week-old puppy she took me out on the ocean and said to me, "Koda, I am going to train you to find whales like I trained your stepsister, Kida." My stepsister, Kida, was 15 years old when I was born. When I was four months old, Kida went to doggy heaven. I then became Mamma Carrie's new whale-watching dog.

Mamma Carrie explained how she trained Kida to find whales and that she would train me the same way. She said, "Koda, your senses are so much better than human senses, and I will use all those incredible senses to find gray whales and other whales, too! I trained Kida by having her see, smell, and hear the whales, and I am going to do the same with you!"

MY SMELL IS 100,000 TIMES BETTER THAN HUMANS
I CAN SMELL UP TO 12 MILES AWAY
EVEN THOUGH I AM COLOR BLIND, I CAN STILL SEE MOTION BETTER THAN HUMANS
I CAN HEAR SOUNDS 4 TIMES FURTHER AWAY THAN HUMANS
IF A HUMAN HEARS A SOUND AT 50-100 FEET THEN I CAN HEAR IT AT 1/4th OF A MILE
SNIFF!
SEE!
HEAR!

My First Day on the Ocean...

Late fall 2016, Mamma Carrie introduced me to a job that I would be doing my whole life. She took me on one of her Zodiac boats out into the nearby Pacific Ocean. Mamma Carrie's boats are ex-Navy SEAL boats, so they are extremely safe, fast, and they turn on a dime. I was a little uneasy at first feeling the powerful waves under the boat, but I soon calmed down because Mamma Carrie said with much excitement, "Thar she blows!" Mamma Carrie saw a whale spouting in the distance. I could not believe how thrilled Mamma Carrie got!

I saw what looked like a puff of smoke, which I soon learned was the whale's spout, also called its blow. I even saw the heart shape. As we approached, Mamma Carrie stopped the boat in the area where the whale blew. The whale had disappeared. Where was the whale?

All of a sudden, I heard a loud WHOOSH! I must admit that the first time I heard and saw the whale that close I jumped into Mamma Carrie's arms. Mamma Carrie said that the blow comes out at 200 miles per hour (mph). Wow, that's fast! When she takes me for car rides, we go about 50 miles per hour so a whale blows four times faster than a car drives! I'm doing math now! That was my first whale encounter ever!

The Migration...

Because it was late October, Mamma Carrie said I would now have to wait until late spring to see our summer resident gray whales return again. She explained to me that during late fall and early winter, about 19,000 gray whales from Alaskan feeding grounds, and another 212 grays from Oregon and other Northwest feeding grounds, head south to the breeding lagoons of Mexico to mate and give birth to their calves. They travel at four mph during migration and it takes them two months to go from Alaska to Mexico, but only about a month from Depoe Bay, Oregon to Mexico.

In the spring, the gray whales head up toward Alaska to feed, however, about 212 of them decide to stay off the northwest coast to eat a different food source. Mamma Carrie told me, "Koda, this small group of 212 whales are the ones I will teach you to identify by their unique coloration patterns and behaviors." I was a little sad because now I had to wait six months before I could see the summer resident gray whales again and begin my job of whale finding!

Destination Depoe Bay!

ALASKA
CANADA
DEPOE BAY,
OR
NEWPORT,
USA
MEXICO

My Training Day...

It is now May and I am nine months old and almost full grown. It is seven o'clock in the morning and Mamma Carrie said, "It is now time to go to work. Today I will teach you how to find the whales and all about gray whale behaviors." Mamma Carrie bought me a bright yellow life jacket that I would always have to wear on the boat to stay safe. She then took me down the stairs to the dock to one of her Zodiac boats. She said, "Up on the boat!" I obeyed by jumping over the tube of the boat and then turning and resting my belly on the tube and draping my legs over the side. We left Mamma Carrie's dock and traveled a short distance to the 50-foot-high bridge, went under the bridge, through the narrow gap and out into the mighty Pacific Ocean. We headed only a quarter of a mile offshore to North Point, a favorite feeding area of the summer resident gray whales.

Mamma Carrie said, "Look! Look! There is a whale coming right toward us and you can see the heart-shaped blow." I was amazed because, sure enough, the blow looked just like a heart. Mamma Carrie mentioned that you can only see the heart shape if the whale is coming directly towards you or is going away from you. The gray whale is a medium-sized baleen whale and due to its size and two blow holes, it forms a heart-shaped blow unlike all other whales.

The Smell...

Mamma Carrie did something I thought was odd. She turned the boat away from the whale and got downwind of the whale's blow. I soon learned why she moved the boat because I got the full-on smell of the blow! What Mamma Carrie didn't tell me was how stinky the breath of a gray whale is...worse than the stinkiest toot ever!

As the whale came closer, I realized how massive this creature was, WOW! Gray whales are so big! I overheard Mamma Carrie saying a gray whale is 40 feet long (the length of a school bus) and weighs about 40 tons. If I do my math right, that's about a ton a foot. I'm only 65 lbs so...80,000 lbs divided by 65 lbs equals 1231 Kodas.

Whale Dive Cycle...

Mamma Carrie said, "Look in the distance, the gray whale is blowing! One blow, now let's time it." It is blowing again and it was only 45 seconds since the first blow. I am waiting 45 seconds and see yet a third blow, then down the whale goes for a deep dive. Mamma Carrie said, "The whale usually blows three times and then stays underwater for three minutes. It can travel up to 400 yards underwater." Wow, that is the length of four football fields!

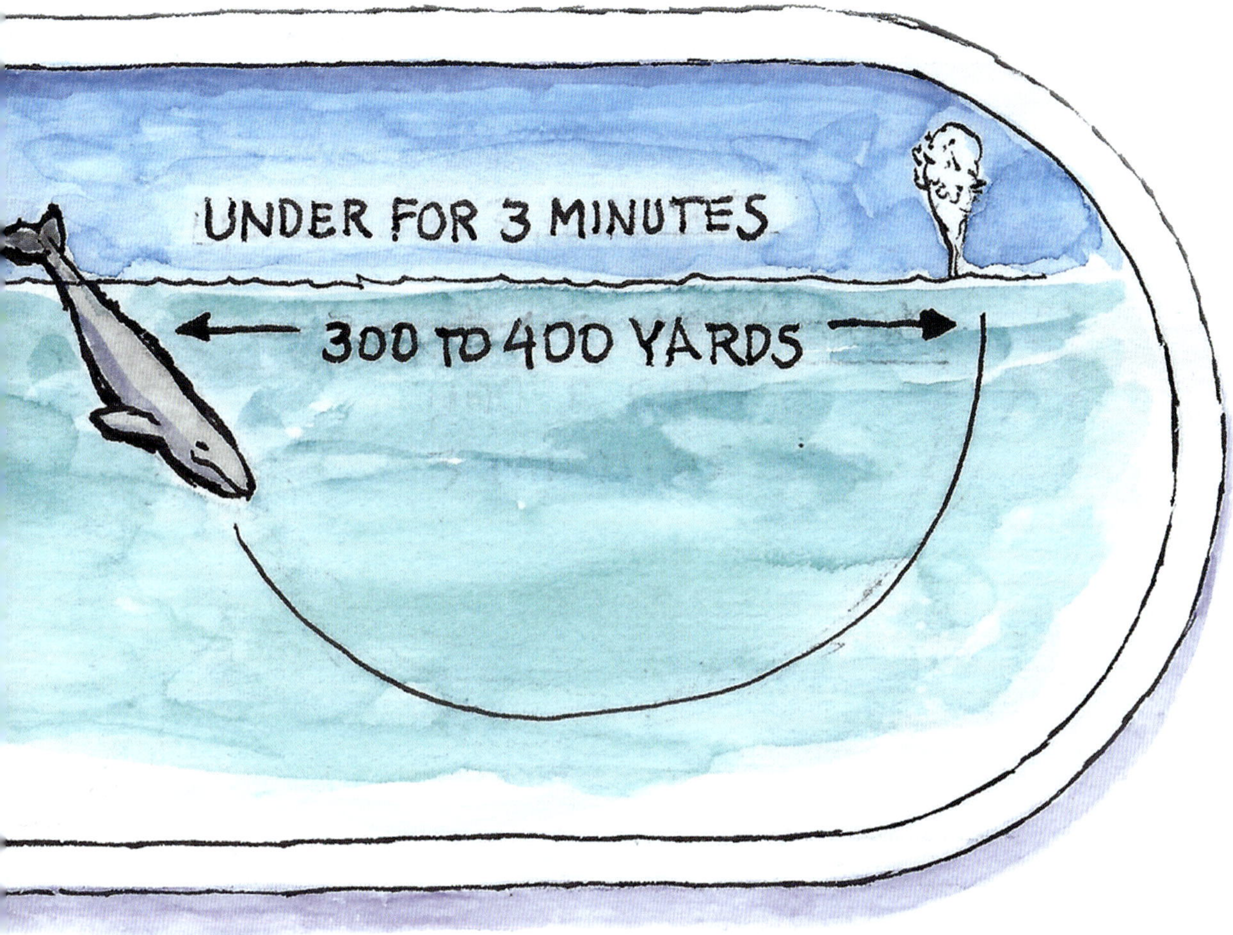

Dive Pattern of a Whale Encounter...

Oh my goodness the whale is coming very close! I am going to bark to let the whale know I'm here. WOOF! WOOF! WOOF! The boat is stopped and all my senses are working. I can smell it, see it, and hear it! Mamma Carrie is saying, "Whale! Whale! Whale!" It blew once, twice, three times. Mamma Carrie said, "Look, you first see the blow, then its back, then the rounded dorsal hump, its pointy knuckles, and finally the ten-foot wide tail fluke." After the whale went under, I saw a large round, smooth area of water that Mamma Carrie called a fluke print. (Turn the page to see this action.)

A Shark...

We waited three minutes...Oh no! There's a shark!!! I see a dorsal fin. Mamma Carrie said, "Oh look, the whale is doing its sharking behavior." Phew, it's just a gray whale feeding on its side, and you can see part of its tail fluke, which looks exactly like a shark fin. Mamma Carrie explained that the mysid shrimp swim just above the bottom of the ocean, which is only about 35 feet deep, and since gray whales are about 40 feet long, the tip of the tail fluke sticks up out of the water as they feed, looking just like a shark fin.

The Discovery...

Why is Mamma Carrie so excited now? She is grabbing a plankton net and trying to collect this red stuff in the water. She explains that the whale just pooped, which looks just like reddish diarrhea, and she needs to use her plankton net to collect it. I learned that Mamma Carrie was the first one to discover what the gray whales off Depoe Bay ate because she collected their poop.

So what exactly is in the whale poo? Mamma Carrie explained that when she was diving in the kelp beds, she saw these ½ inch shrimp-like creatures called mysid shrimp. She thought that's what the whales were eating, but she had to prove it. What better way to know what an animal eats then to collect its poop? She also dropped the net to the bottom to collect the live mysid shrimp.

When she pulled up the net and I looked into it, I was shocked because the food was the size of my toenail! Mamma Carrie said that the gray whales eat a ton a day. That's 2000 lbs of these ½ inch mysid shrimp each and every day.

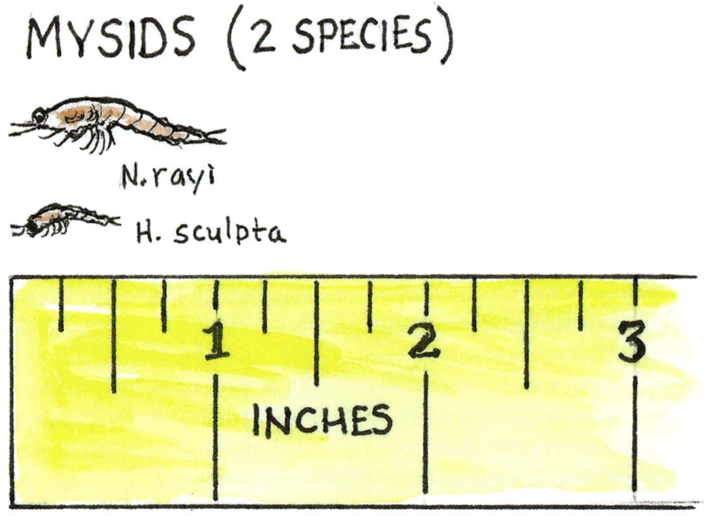

CARRIE'S FINGERTIP AND MYSID
KODA'S PAW AND MYSID

How Baleen Whales Feed...

I was curious about how an animal so big collects and eats these small creatures. Well, Mamma Carrie must have read my mind. She said they usually feed on their right side but sometimes there are lefties who feed on their left side. Everyone is unique, even whales. To get the food, they suck up the shrimp from just above the bottom of the ocean. They then filter the shrimp and seawater through their 300 blonde baleen plates that hang down from their top jaw. The baleen plates trap the food, but the water flows through the openings of the plates. After the food is trapped, the whales wipe off the baleen with their two-ton tongue. Wow! That's a big tongue!

Have you ever tried to clean your teeth after eating peanut butter? I love peanut butter! Well, that's what a whale does when it cleans off the mysid shrimp from its baleen. What surprises me even more is that Mamma Carrie said a gray whale's throat is only the size of a grapefruit! I guess I never have to worry about being swallowed by a gray whale. Not that I was ever worried by the way! Wink.

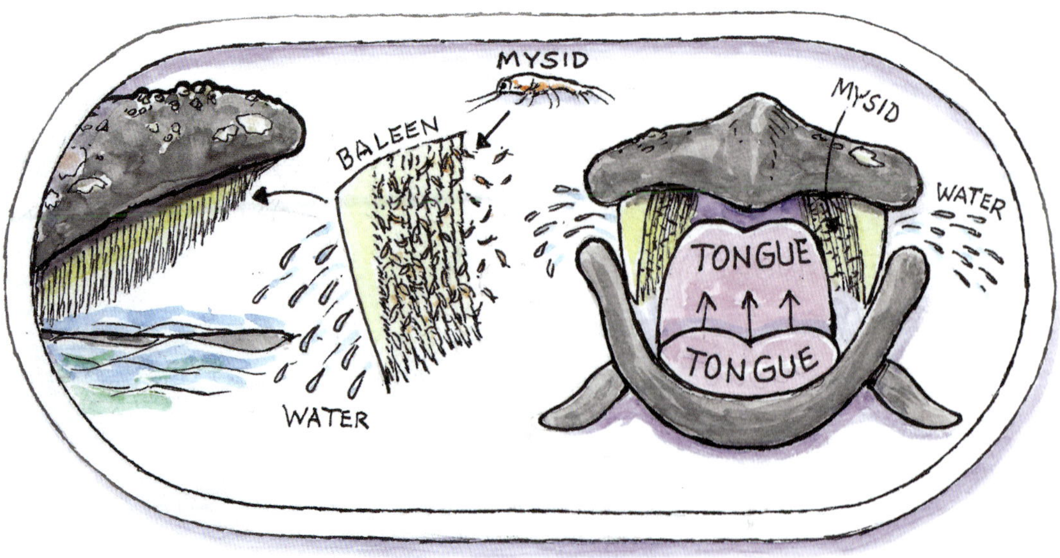

The Kelp Beds...

After Mamma Carrie collected the poop, she got on her SCUBA gear, grabbed her plankton net, and dove to the bottom of the ocean, which is only about 35 feet deep in Depoe Bay. Mamma Carrie wanted to collect the food from the same area she collected the poop. She explained that the mysid shrimp live right above the rocky bottom near the holdfasts (root-like structures) of the bullwhip kelp. To collect them, you have to swim right above the bottom where they are found. The swarms can be two feet to 18 feet thick above the ocean bottom.

Mamma Carrie explained to me that the bullwhip kelp makes a unique underwater forest that is home to rockfish, salmon, sea stars, sea urchins, sea anemones, and many other marine animals. The kelp can grow to be 100 feet long and can grow a foot per day. That's incredible! I wish I could dive underwater like Mamma Carrie and see all those awesome critters, too!

ORANGE SEA STAR

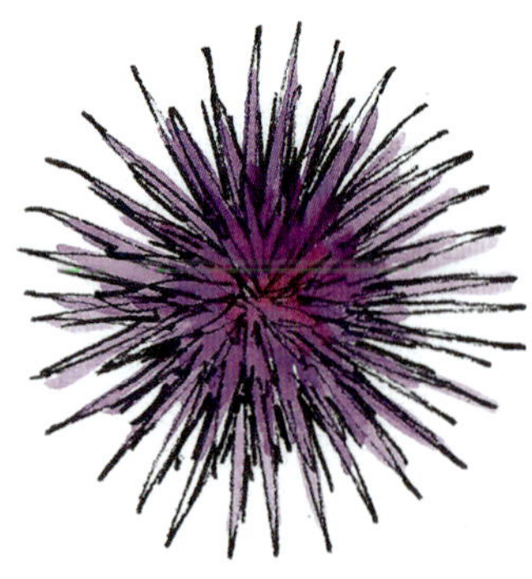

PURPLE SEA URCHIN

The Spyhop...

I waited three minutes! Where will the whale surface this time? I think I hear something. Suddenly, an eight-foot-tall head rises vertically out of the water right next to the boat and I saw its eye. I was shaking with excitement. The big, brown eye was the size of an orange! As it stared into my own brown eyes, our gazes met, and we looked into one another's eyes for what seemed hours, but was in fact only seconds. Then the whale silently slid straight back down into the water. Mamma Carrie said, "Koda, you got to see your first spyhop! A spyhop is where a whale comes vertically out of the water to its eye level. You are so lucky to see this whale behavior!"

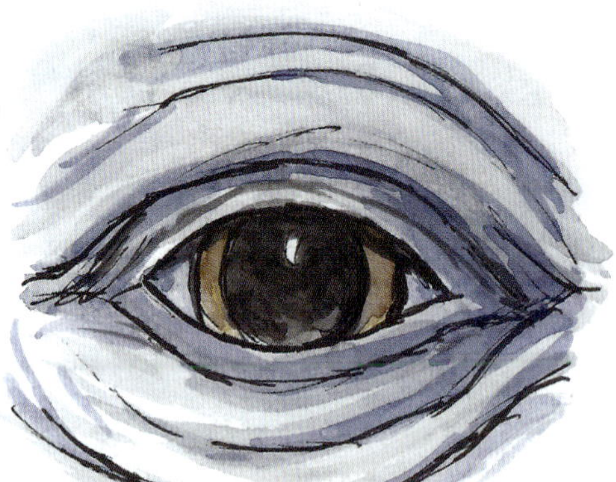

GRAY WHALE'S EYE

ORANGE

The Breach...

We waited another three minutes... All of a sudden, Mamma Carrie started squealing, I looked to my left and saw the whole body of a gray whale coming out of the water and making a big splash as it returned back into the sea. Mamma Carrie is bouncing with excitement, I'm barking and the whale is repeatedly jumping out of the water, once, twice, three times, four, five, six, seven, and finally eight times! After it stopped, Mamma Carrie said, still shaking with excitement, "Koda you just saw a whale breaching, where at least three quarters of its body comes out of the water." After all that excitement, we headed back to the harbor. It was a fun, thrilling and educational day! I am so tired!

Nap Time...

As I lay on my doggy bed, I thought about my first official day on the ocean. I sure learned a lot! I learned that gray whales are about 40 feet long and weigh 40 tons, about a ton a foot, grays stay underwater for three minutes, they eat mysid shrimp, their poo looks like reddish diarrhea, they have a heart-shaped blow, a fluke print, and stinky breath! The awesome behaviors I saw were fluking, sharking, spyhopping, and breaching!

Mamma Carrie is such a good teacher. She taught marine biology at colleges for 35 years! I overheard Mamma Carrie say that I am going to be a great whale-finding dog just like my stepsister, Kida. She also said I must be her lucky charm, because I already had so many incredible whale encounters. I am going to have Mamma Carrie tell me more stories about my stepsister, Kida...after I take a nap. The ocean really tired me out!

GRAY WHALE BEHAVIORS
BLOW
HUMPING UP
FLUKING
BREACHING
SPYHOP
SHARKING

Yogi...

I was up at seven o'clock and ready to go out on the boat again. Mamma Carrie said, "We will learn how to identify individual gray whales today." We went down the dock stairs and onto her Zodiac boat. We went under the big bridge and out into the bay. We headed to North Point where whales love to feed. Mamma Carrie stopped the engines, and we waited. Oh my goodness, a whale is coming straight at us. Mamma Carrie said, "Here comes Yogi. Yogi has a bear face on her left dorsal hump. Yogi was born in 2014 and is a female. You can identify individual gray whales by the patterns on their dorsal hump. Koda look, Yogi is in the kelp bed, sharking."

Does Yogi see me? WOOF WOOF! There is her blow! And YES! She is coming right at me! She is so close I can see the bear face on her dorsal hump as she humps up and now the tail fluke comes up and lastly, the smooth circle of the fluke print. Wow, that was so awesome! Mamma Carrie took photographs of Yogi's dorsal hump for her research. She writes papers and books on gray whales.

Rambolina...

Mamma Carrie said, "When finding whales, you need to know your directions on a compass! Now we will head north so the bow (front) of the boat is north on the compass and the stern (back) of the boat is south. If the whale is to the left of the bow, going north, then it's northwest but if it's right of the bow going north then it's northeast. Does that make sense, Koda?"

Yes it does, and I smell something coming from the northwest. Now I see the blow. I start barking to let Mamma Carrie know that there is a whale. We travel to Pirate's Cove where we see a very large blow. Mamma Carrie said that female gray whales have higher blows (10' to 12') than male gray whales (6' to 8'). I bet the whale up ahead is a female. Wow, she is very large! It looks like she has parallel marks on her left side. Mamma Carrie said her name is Rambolina X Proper for the 10 prop cuts she got from being hit by a boat propeller in the 1980s. She is very old, at least in her 50s.

Comet and O'Valentine...

Wow, I see more large blows south near the buoy. Look, sea lions are resting on the buoy. Now I see a weird, lopsided heart blow, and Mamma Carrie said, "Oh, it's Comet! She has a blockage in her left blow hole and that makes her blow lopsided." Mamma Carrie said that many times Comet comes right next to the boat and shows off her beautiful black tail fluke. She has been so friendly over the years, that Mamma Carrie named one of her boats after her.

One of Comet's best friends is O'Valentine. She is named O'Valentine because she has a heart shape on the underside of her right tail fluke. O'Valentine also has an "O" spot on her right dorsal hump. Look how they travel side-by-side! They are such good friends!

BUOY WITH SEA LIONS

Eagle Eye...

I looked to my right and saw yet another whale, but it has a shorter blow. It must be a male. Mamma Carrie glanced over and excitedly said, "My honey, Eagle Eye, I love you!" What I learned is that Eagle Eye is Mamma Carrie's favorite whale and that if she was a gray whale, Eagle Eye would be her boyfriend.

He came right up to us, and I swear he winked at Mamma Carrie. I can see why Mamma Carrie likes him so much because he is beautiful with his dark color and unique "Eagle Eye" pattern on his left dorsal hump. Mamma Carrie said he was born in 1994 to Snake Eyes. She also said he comes almost every year to Depoe Bay and checks out the food, and females. Mamma Carrie hopes he comes to see her too! She named one of her boats after Eagle Eye.

Story of Eagle Eye's Close Encounter with Kida...

Mamma Carrie shared a story about Kida and Eagle Eye. "I have loved Eagle Eye since I first met him in 2004. At that time Eagle Eye was ten years old and would swim towards the boat. Kida was three years old and had just started to learn her right from her left. Kida's newest behavior was to put her right ear up if the whale was coming from the right and her left ear up if the whale was coming from the left. Eagle Eye helped Kida practice her right from her left because first Eagle Eye swam towards the boat from the right and up went Kida's right ear and then Eagle Eye swam up from the left and up went Kida's left ear. Eagle Eye helped Kida practice her right from her left for over an hour that day. I really think Eagle Eye and Kida enjoyed each other's company!" I learned on a boat, the right side is called the starboard and left side is called the port side (remember four letters in left and port). Also you can remember which is your left if you put your left hand up, or in my case paw, so your pointer finger or toe and thumb make a letter L, for left.

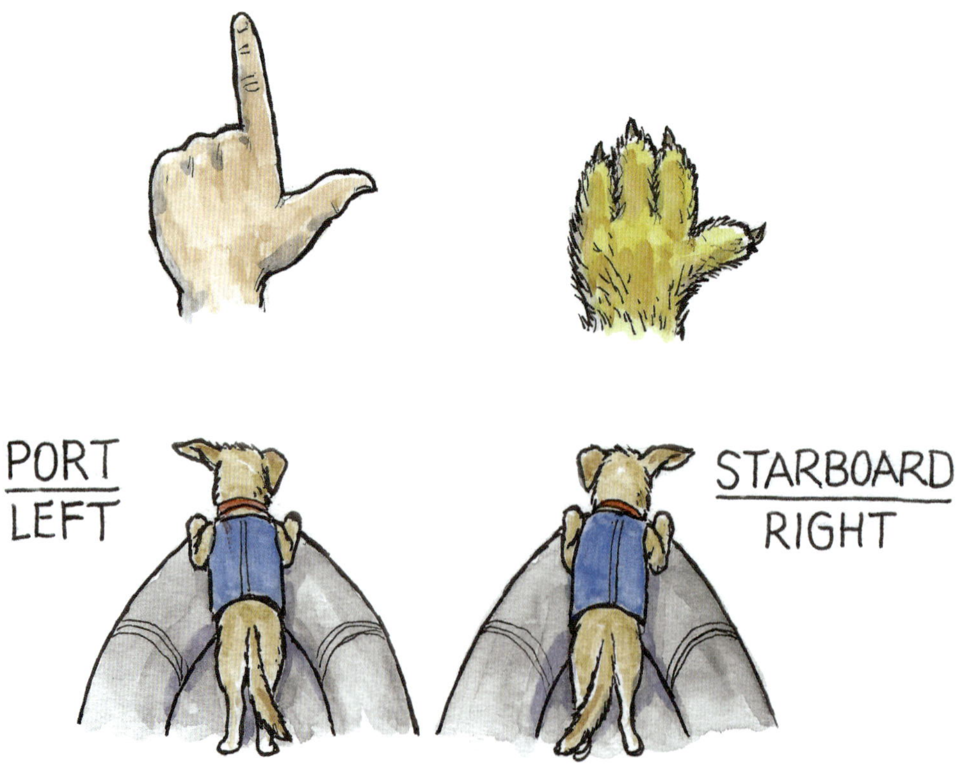

Morisa...

We saw another blow further south. Mamma Carrie said, "Oh, look! Morisa showed up to her favorite feeding spot at South Point. Morisa is an easy whale to identify because she has a dot and a dash on her left side like Morse code. The dot and dash in Morse code means "A." At first, because of the shorter blow, we thought we had a Moris but now we know she's a Morisa. Since she had a dot/dash (= A) we added A to Moris to become Morisa."

I wondered about the dot/dash code Mamma Carrie was talking about. She said, "Koda, before you or I were born, a form of communication was invented in the 1830s using dots and dashes for letters and numbers. This was called Morse Code. It was used during World War II to send important messages."

"Morisa has brought many calves to Depoe Bay. Since she taught her calves where to feed, they also now come to Depoe Bay to feed. Morisa can also be identified by a crescent moon shape on her right side plus a bite mark on each side of her tail fluke where killer whales tried to stop her when they attacked her. She is a survivor! Morisa also does a unique behavior called bubble blasting where she blows bubbles in a circle to either trap prey or for buoyancy (letting air out to sink down underwater)."

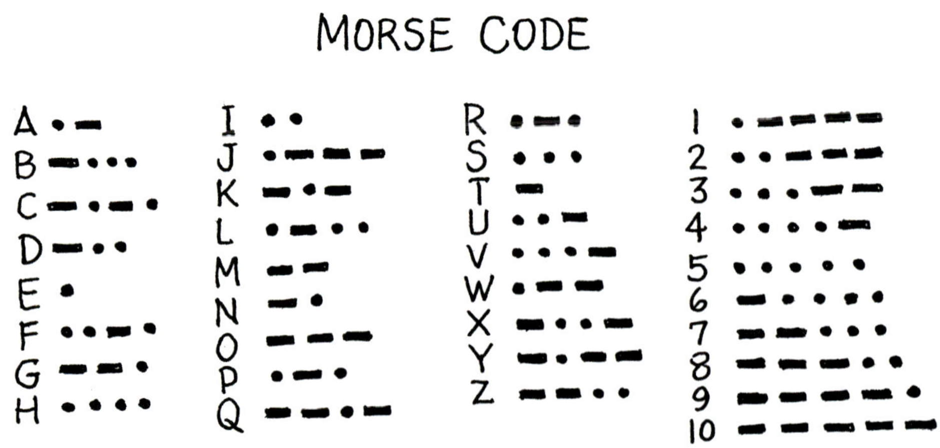

Ginger...

In the distance, I saw the blow of yet another whale. This whale had a tall, black tail that rose high out of the water! Mamma Carrie said, "Oh, it's Ginger, and she is showing off her beautiful black tail fluke again! She shows off her tail more than any other whale I have ever seen!" When we got closer, I noticed that Ginger had an all orange head. Mamma Carrie told me that the orange was whale lice but not bad lice because these orange whale lice help in areas that have cuts by eating up the bad skin. I also noticed cuts on her side and wondered how she got them.

Mamma Carrie was reading my mind and said, "Look at the boat propeller cuts on her side. She got hit by a boat in June of 2016. Luckily the cuts are not deep, so she should heal up fine. Ginger was born in 2010, so she was six years old when she got hit by the boat propeller."

Look, I think Ginger is winking at me! I love Ginger! I am beginning to love whales as much as Mamma Carrie does.

Blanco...

Mamma Carrie got a phone call that her five grandkids had come to see her. We left the whales and went back into the harbor. After the boat was docked, Mamma Carrie got out and gave each of her grandkids a hug and kiss. I can tell she loves them very much. She hopes that one of her grandkids will become a marine biologist just like her.

The grandkids got on the Zodiac boat and brought Kida's blue life jacket for me to wear. We went out of the harbor, under the bridge, and south towards Gull Rock named for all the nesting gulls on it. All of a sudden a whale appeared out of nowhere. It looked like a ghost because it was all white. Mamma Carrie said, "Oh my goodness, it's Blanco. Blanco means white in Spanish, and he is very white." He came closer and closer. I was barking, the grandkids were yelling with excitement, and Mamma Carrie kept saying, "Blanco we love you!"

Blanco came so close to the boat that I almost touched noses with him. He circled the boat, playing with us and then he would blow until we all were soaked with whale blow breath. Once again he stuck his nose next to the boat and I began shaking with excitement. This is my best whale encounter yet! Mamma Carrie was so happy that her grandkids, Kai, Kale, Luke, Zach, and Ellie got to experience this absolutely incredible whale encounter. Mamma Carrie said it was time to go so we went back to the dock. After we got back, she hugged her grandkids goodbye, and then I went upstairs to lay on my bed. Mamma Carrie said she would tell me the story of Kida and Blanco.

Story of Blanco's Close Encounter with Kida...

I was lying on my bed, partially dozing after a long, exciting day on the ocean with Blanco and all the other different whales when Mamma Carrie came in and said she wanted to share a story about Kida. Mamma Carrie said, "Kida loved her gray whale family! Some of her favorite whales were Blanco, Yogi, Scarback, Eagle Eye, Comet, Ginger, Morisa, Rambolina, and O'Valentine. Blanco was probably Kida's most favorite whale because Blanco and Kida would always tease each other. Kida met Blanco when she was 11 years old. He seemed to be an outcast, with none of the other whales wanting to interact with him, so he must have decided human and dog companionship would be fun! Many times over the next three years he would come up to the boat on Kida's side and play with her by blowing in her face, spyhopping next to the boat, playing hide and seek or just trying to touch her nose."

"One day Blanco spyhopped next to the boat and blew so hard that Kida's ears flew backwards, and she smelled like stinky whale breath. Another time Blanco tried to sneak up to the boat and scare us with his loud blow but what he probably didn't realize was that because he was so white we could see him coming through the water. Kida and Blanco had a special connection! Koda, I am so glad you had that awesome encounter with Blanco today, too! Blanco is such an incedible whale!"

Scarback...

The next morning we had just traveled into the bay when Mamma Carrie got a call on her radio that her five grandkids were on the dock and wanted to go see whales again. She rushed back to the harbor and picked them up. As she was heading under the bridge, she got another call that the most famous summer resident gray whale, Scarback, had just showed up off Whale Cove.

We sped down to Whale Cove. I could see Mamma Carrie shaking with excitement! She said, "Kids, look, do you see Scarback, our most famous whale? Her name is Scarback because she got hit by an exploding harpoon back in the 1980s. An area over eight feet wide was blown away by the explosion. It is filled with orange whale lice that eat the bad skin and probably keep it infection free. I have seen Scarback almost every year since 1992 and the wound has not changed. She has brought many calves to Depoe Bay and her calves have returned to feed here too. An observation that I made is that as Scarback ages, her tail is turning from dark gray to white just like people getting white hair when they are old. Look, kids, here she comes, and it looks like she is going to give us a close encounter! She is such a friendly whale!"

I wore Kida's blue life jacket again and it must have been lucky because we had many amazing interactions with Scarback. We returned to the harbor. Mamma Carrie said when we get home she would tell me another story about Kida that was both happy and sad.

Story of Scarback's Close Enounter with Kida...

We went upstairs and Mamma Carrie said she would tell me Kida's story with Scarback. Mamma Carrie said, "Koda, you have had amazing whale encounters this summer! Over the years so did your stepsister, Kida. I was so glad that Blanco and Scarback gave you the awesome interactions that they also gave Kida when she was my whale-finding dog. I have one last story to share about Kida. In October of 2016, I took Kida on her last ocean trip to see whales. She had become sick with cancer and soon would leave this earth. We were at Whale Cove when Scarback came right up to the boat, and she raised her beautiful white tail fluke right next to where Kida was hanging over the side. I think Scarback knew that Kida was really sick because Scarback came and fluked next to Kida over a dozen times. Whenever I moved the boat, Scarback would follow us. Just before I left to go back to the dock, Kida and Scarback's eyes met and they seemed to say a tender goodbye. This would be Kida's last whale encounter."

Mamma Carrie has me in tears now even though dogs are not supposed to cry. I wish I could have gotten to know my stepsister Kida better! We miss Kida! Mamma Carrie said, "That's all for tonight. Koda, I want to leave you with a happy thought! Tomorrow I have a special surprise for you! Dream well tonight!"

The Surprise...

The next morning I was up super early and wondering what surprise Mamma Carrie had in store for me. I ran down the stairs to the dock and there before me was a brand new Zodiac boat. It was a beautiful royal blue color, my favorite color by the way. Mamma Carrie said, "Koda this is my new boat and I am naming it Krazy Koda after you." I was thinking to myself, I am not crazy, but the more I thought about it, I am "Krazy" for whales. Mamma Carrie had painted on the cabin of the boat my picture and the name *Krazy Koda*. Mamma Carrie said, "The boat is 24 feet long, that is six Kodas lined up head to tail." I divided 24 by six and figured out that I am four feet long. I'm doing math again! I jumped up and down many times because I was so happy I had a boat named after me! I hopped on my boat and we went for our maiden voyage.

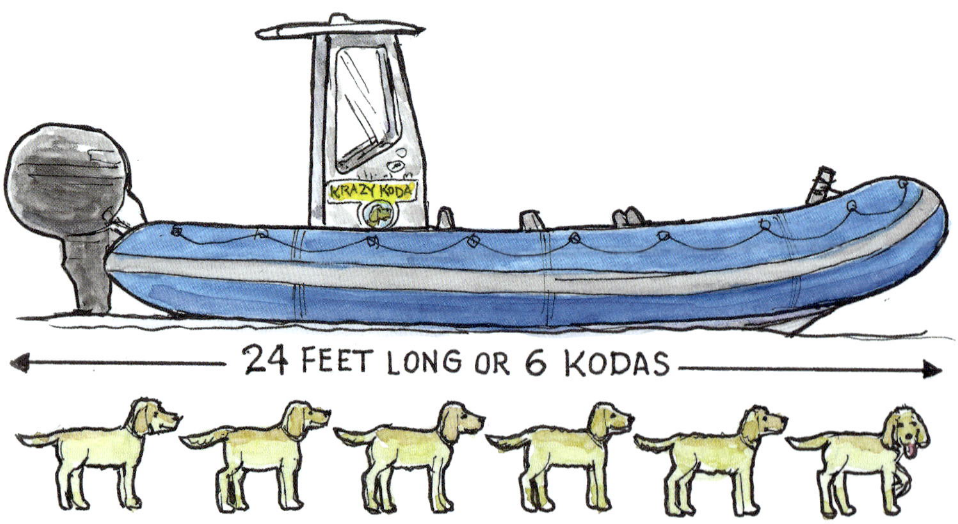

KRAZY KODA

Under the Bridge and Back Again...

We went under the arch of the 50-foot-high bridge and into the open ocean off Depoe Bay. I glanced around at all the whale blows I was seeing and recalled all the information Mamma Carrie had taught me this summer. I knew how to tell a male from a female blow since female's blows are higher and I knew the behaviors of breaching, spyhopping, fluking, sharking, and bubble blasting. I learned that Mamma Carrie was the one to discover that gray whales off Depoe Bay ate mysid shrimp and this is why the gray whales stay here in the summer. She proved they eat mysid shrimp by collecting whale poop. I also unfortunately learned that whale breath is very stinky! I also know how to identify individual gray whales including, Scarback, Eagle Eye, Yogi, Rambolina, Ginger, Morisa, Comet, Blanco, and O'Valentine. Whales are individuals just like you and me! They are different individuals, but the same inside! I am so lucky to have such an awesome job! I am the whale-watching dog of Depoe Bay!

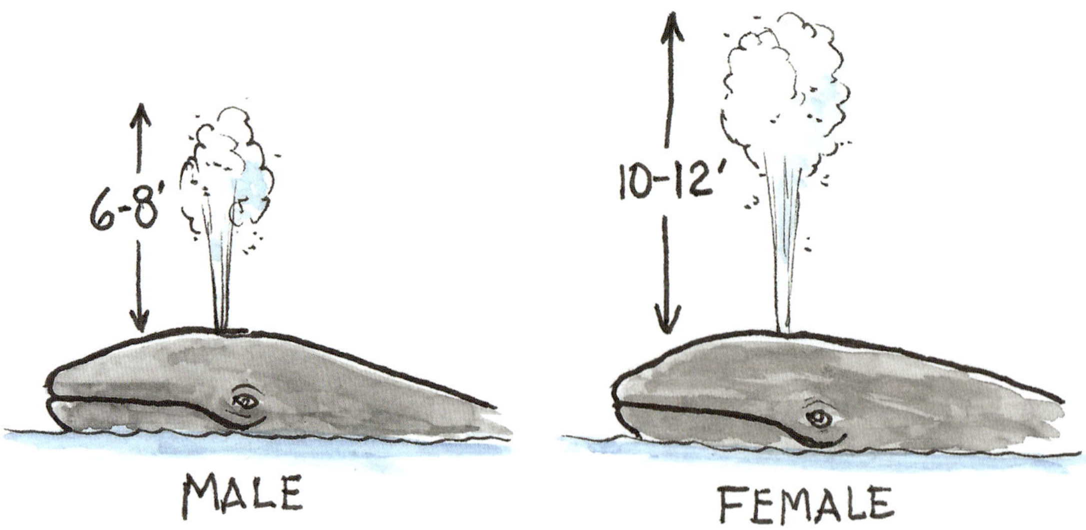

A Tender Goodbye...

Mamma Carrie and I went on one last ride at the end of the summer and saw something spectacular! Mamma Carrie showed me that early in the morning or late in the day when the sun is out, you can see what she calls a rainblow where all the colors of the rainbow are visible in the blow of the whale. Mamma Carrie said to remember the order of the rainbow colors, memorize ROY G. BIV with each letter being the first letter of the color: red, orange, yellow, green, blue, indigo, and violet.

On our very last whale sighting for that trip, we saw another rainblow, and as I stared above the rainbow of the blow, I saw a cloud that looked just like Kida with her right ear up and running over the rainblow of the whale. Kida was going over the rainbow bridge. A tear formed in my right eye as Kida was giving us a last tender goodbye.

The Next Generation...

Mamma Carrie has a fleet of five boats named either after her dogs, Kida and me, or her special whales, Comet, Eagle Eye, and Sea Star. Here is her fleet with her biggest boat and oldest grandson to her smallest boat and youngest granddaughter. At the helm (steering wheel) are Kai on Comet, Kale on Kida Girl, Luke on Eagle Eye, Zach on Krazy Koda, and Ellie on Sea Star. Mamma Carrie and I are training my daughter Kilo to be a new whale-finding dog. I named her Kilo because she was 2.2 pounds in 2.2 weeks and a kilogram equals 2.2 pounds. Mamma Carrie and I are trying to teach her how to find whales like my stepsister Kida did and I am doing now. Wish us luck in training her!

What will Grandma Carrie's grandkids learn about the gray whales? Will they make new discoveries, too? I say YES! Will Kilo have a puppy of her own to train to become a whale-watching dog? I was born in 2016 and now am eight years old in dog years, but in human years I am 56 since one dog year equals seven human years. Kida lived to be 15 dog years. I hope I can do whale watching for over 100 human years like my stepsister Kida did. I wonder what the future holds?

Glossary of Whale Words

Arthropods: a group of animals that have a hard outer covering called an exoskeleton that they molt (examples are insects, crustaceans, spiders, and others)

Baleen: a fingernail-like material, keratin, that hangs down from the top jaw of a baleen whale

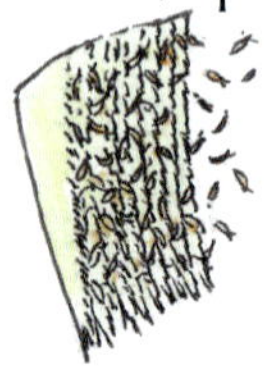

Baleen Whale: a whale that has two blow holes and feeds with baleen on tiny organisms
Barnacles: an arthropod that lives on its head and filter feeds with its legs, covered with plates
Blow: (aka Spout) the breath of a whale combined with water vapor
Bow: front of a boat

Breach: behavior when 3/4 of the body of a whale comes out of the water and then the whale lands on its side

Bullwhip Kelp: a brown seaweed that can grow 100 feet long and grow one foot per day
Crustacean: a group of aquatic arthropods that usually have their head and thorax fused together (examples crabs, shrimp, lobsters, mysid shrimp, amphipods, and others)

Dorsal Hump: a body part of a gray whale is the hump found about 3/4 of the way along the back in the same place that a dorsal fin would be on other whales

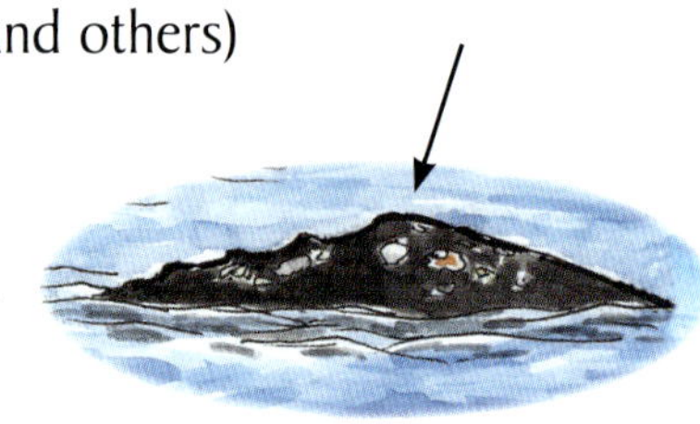

Echinoderms: a group of sea organisms that has five-sided radial symmetry and moves with tube feet (examples are sea stars, sea urchins, sand dollars, and others)

Fluke: the tail of a whale used for locomotion

Helm: Steering area
Holdfast: root-like structures holding seaweeds to bottom but don't absorb water or nutrients
Kelp: a type of long (greater than one foot), brown seaweed found in cool waters
Migration: going from one location to another
Kilogram: a metric unit of mass where one kilogram is equal to 2.2 pounds

Knuckles: a body part of a gray whale which are the pointy bumps after the dorsal hump

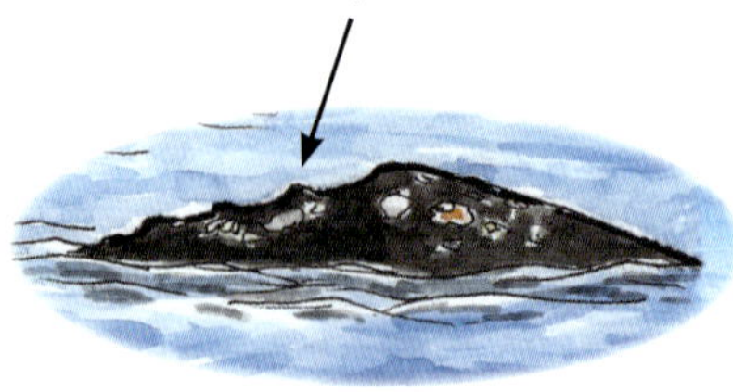

Mysid Shrimp: a unique type of crustacean that is ½ " to 1" long and found in large groups

Plankton Net: a net with very small holes to collect small sea creatures

Port: the left side of a boat

Propeller: the part of a boat that pushes a boat through the water

Rainblow: a term invented by marine biologist, Carrie Newell, where the colors of the rainbow are seen in the blow of a whale

SCUBA: Self Contained Underwater Breathing Apparatus used by divers to breathe underwater

Sea Star: an animal (echinoderm) with usually five arms and tube feet (incorrect to call it a starfish because it is NOT a fish)

Sea Urchin: an animal (echinoderm) with spines coming out of a structure called a test

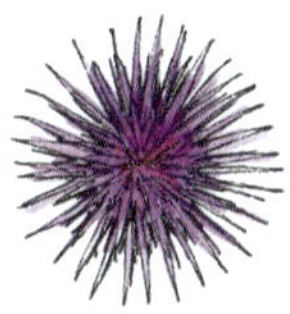

Sharking: behavior invented by marine biologist, Carrie Newell, where the tip of the tail fluke of a gray whale sticks above the surface and looks like a shark fin

Spyhop: behavior when a whale rises vertically up out of the water just above its eye

Starboard: right side of the boat

Stern: rear of a boat

Toothed Whale: a whale with one blow hole and feeds with teeth on larger organisms

Ton: 1 ton = 2000 lbs (pounds)

Whale Lice: an orange crustacean, a type of amphipod, that lives on a whale and eats dead skin

Zodiac: a type of boat with a hard bottom and rubber-like tubes

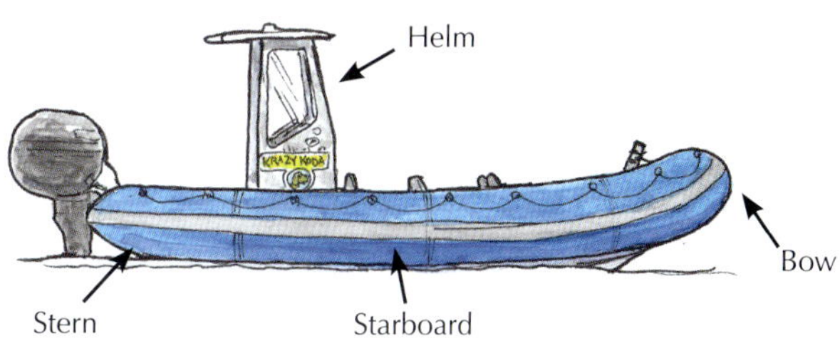

What did you Learn about Whales?

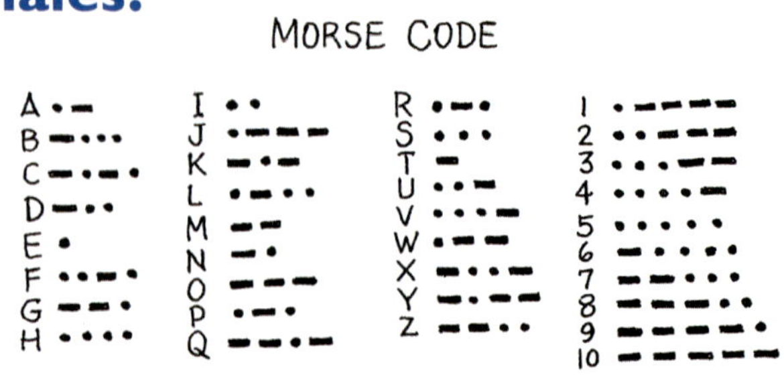

1. Write your name and age in Morse Code.

Name is ______________________

Age is ____________

2. Name four (4) behaviors that a gray whale performs.

A. _________ B. _________ C. _________ D. _________

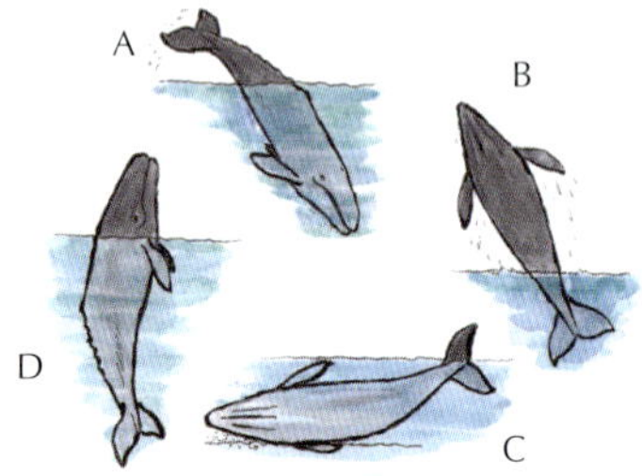

3. After a whale blows, what parts of its body do you see?

A. _________ B. _________ C. _________

and finally it leaves a smooth round circle called a D._________.

4. Before a deep dive, how many blows? _________

How many seconds between blows? _________

How long is a gray whale's deep dive? _________

How far can it swim in that amount of time? _________

5. How many mph does the blow come out at? _________

6. What shape is the blow of a gray whale if it comes towards you or away from you? _________

7. a. In the migration route of a gray whale, what country do they go from where they eat and then what country do they go to where they have their babies (calves)?

Go from feeding country of _________ to calving country of _________

b. How many mph do gray whales travel at when migrating? _________

8. What three (3) senses do dogs use to find whales? A._________ B. _________ C. _________

9. How many times better is a dog's sense of smell than a human's sense of smell? _________

10. Dogs can hear sounds _________ times farther away than humans.

11. Draw a circle around the letter of the compass where the sun sets, and draw a square around the letter of the compass where the sun rises. Now if a whale is traveling NE, draw an arrow on the compass for that direction and if a whale was traveling SW, draw an arrow on the compass for that direction.

12. How many pounds (lbs) is a ton? __________

13. How many Kodas would equal a 20-ton whale? __________

14. Explain how baleen whales feed. They first __________ their tongue then they __________ the food and __________ the food in their __________ plates and wipe it off with their __________ ton tongue and swallow it in a __________-sized throat.
(Use these words for the blanks: trap, grapefruit, depress, baleen, two, suck up)

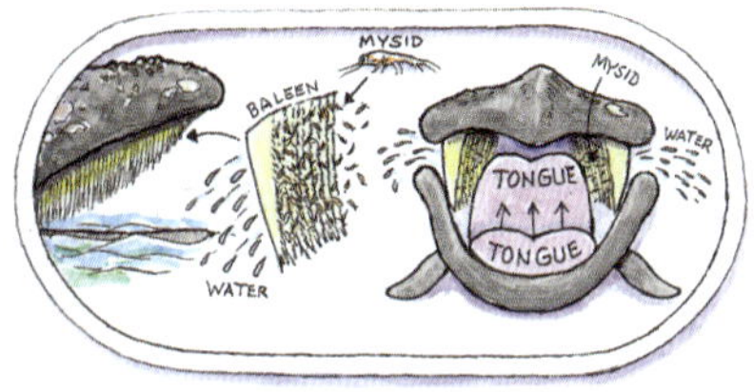

15. How long can bullwhip kelp grow to be? __________
How much can this kelp grow in a day?__________

16. A gray whale's eyes are the size of an __________ and color of the eye is __________?

17. From the pictures A-D, give me the name of the whale and one unique fact.

 A.________
____

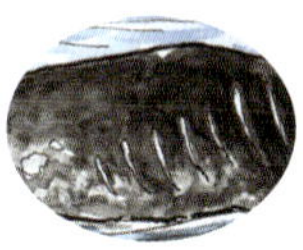 B.________
____

 C.________
____

 D.________
____

18. List the colors of the rainbow in order. Remember about Roy G.Biv.

__________, __________, __________, __________, __________, __________, __________

19. If the boat Krazy Koda is 24 feet long, then how long is Koda? __________

20. How can you tell the difference between a male and female gray whale when looking at their blow? A male's blow goes________ feet high whereas a female's blow goes __________feet high.

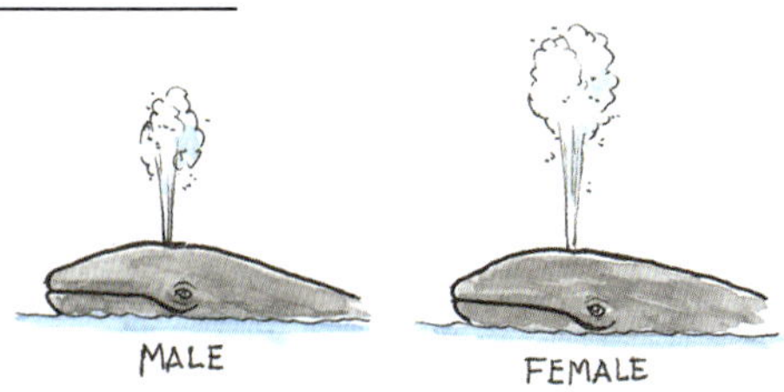

Author Biography

Carrie and Kida

Carrie and Koda

As a four-year-old child, my dream was to become a marine biologist. I grew up in Michigan and moved to South Dakota, receiving a BS degree in Fisheries and Wildlife and then I moved to Utah receiving a Composite BS degree in Zoology, Botany, and Geology. My next move was to Arizona with a graduate degree in Invertebrate Zoology. This degree opened a job at a college in Eugene, Oregon. My two girls and I moved in 1992 where I began teaching Marine Biology. I worked with gray whales on weekends. I received my last graduate degree in Biological Oceanography since I was the scientist to discover what the gray whales ate off the Oregon coast. After my discovery, Jean Michel Cousteau featured my research in his 2006 PBS documentary *Gray Whale Obstacle Course*. Another lifelong dream to dive with the Cousteau divers came true! My research was also featured on an Oregon Field Guide episode. After 35 years, I retired from being a Marine Biology professor, but I am still active in doing gray whale research. I also have run a whale-watching business for almost 20 years called Whale Research EcoExcursions and I have a Whale, Sealife, and Shark Museum. I captain trips for people to see whales with Koda at my side helping me. This book is the true story of how gray whales have interacted with my dogs.The whales seem to know and love the dogs and I believe this is why many times on our whale-watching trips these whales give us such close encounters. You can see more about the tours, books, and museum at www.oregonwhales.com.

Illustrator Biography

Jack Wiens has illustrated numerous children's books for other authors and recently illustrated and wrote his own children's book for all ages named, *What Bear Said about Life, Love, & Other Stuff*. Also from Jack, a little book about loss, *Tending Our Grief*, available on Amazon. He lives in Ashland, Oregon, and loves coming to Depoe Bay to whale watch with Carrie and Koda. You can see more of his work at www.jackwiens.com.